The Twenty-First Century New Way of Seeing

The Twenty-First Century New Way of Seeing

A Nexus of Art, Science, and Ethics

Elta Marie Wilson

WIPF & STOCK · Eugene, Oregon

THE TWENTY-FIRST CENTURY NEW WAY OF SEEING
A Nexus of Art, Science, and Ethics

Wipf & Stock
An Imprint of Wipf and Stock Publishers
199 W. 8th Ave., Suite 3
Eugene, OR 97401

www.wipfandstock.com

PAPERBACK ISBN: 979-8-3852-7081-1
HARDCOVER ISBN: 979-8-3852-7082-8
EBOOK ISBN: 979-8-3852-7083-5

VERSION NUMBER 02/23/26

Soli Deo Gloria

Acknowledging my mentors with thanks:
The Very Rev. Ian Markham, PhD, president and dean of the Virginia Theological and General Seminaries;
Dr. Aaron Rosen, PhD, Executive Director Clemente Course, Humanities, visiting professor, King's College, London, founder of Parsonage Gallery, Maine, United States;
Boston School of Fine Arts; and
the Art Department of Fort Lewis College, Durango, Colorado.

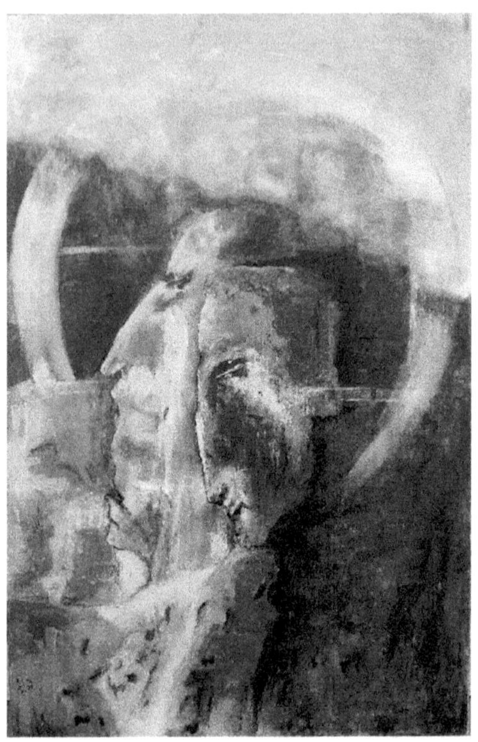

Both artists and scientists are visionaries.
"Rock Icons: Arches Nat'l Park, 6 mi. marker", by Elta Wilson, acrylic on canvas, 24" x 36", 2009.

Fully abstracted art numbs the viewer as if reaching for oblivion with a martini or a toke. Art in the twenty-first century must be grounded in science, substance, and ethical inquiry. Millions paid for exploding bananas and stuffed plastic sharks is obscene. Total abstraction fails because it avoids the tough questions: Where do we come from? What are we? Where are we going? What is this art about and why? Within the Scientific Revolution of the twenty-first century, both artists and scientists are visionaries offering substance to guide our future.

—Elta Marie Wilson, JD, DMin—artist, theologian, journalist

Contents

Introduction

Where Do We Come From? What Are We? Where Are We Going?

We live on the cusp of the Scientific Revolution and welcome the golden age of astronomy that impacts us with knowledge of billions of galaxies. Yet our perspective is grounded throughout the ages with earthbound philosophies, ethics, and religious notions that constitute memories without a path to the future of space and understanding the reality of our universe and its exploration. There seems no means of comprehending our true size. Our own entire solar system is too small to pinpoint within the Milky Way.

A mere four-hundred-plus years ago in 1600 CE Dominican Friar Giordano Bruno was burned at the stake for insisting that the earth moved around the sun. Where will science find us in the next four hundred years? Will we find and learn of the values of our true size? Will we understand the notion of time? Does time weave in and out like cloth, expanding and slowing with the design of the universe or multiverse? Will our understanding of death and perhaps "eternity" expand to encompass the universe itself?

Are we so small that we do not count? Or do we? How do we understand true size? We might visit the Grand Canyon and feel the land. With rocks over 30 million years old surrounding us, we could breathe in the age of the earth and find a split second of connection. But how do we understand a universe that is so vast we do not even know its true size or shape or age? Are we so limited that we cannot even imagine it? We are sensate beings with an understanding of our size on earth, but how do we expand our psyches to embrace our new knowledge of the universe?

As we face the most confounding issues in human history, artists draw designs worthy of any bathroom. Abstraction, they call it. Art connoisseurs boast, "Wonderful colors! And look at those painterly brushstrokes!" I ask, "So what?"

Art for art's sake is dead. It lacks fundamental substance for the twenty-first century. It is self-absorbed. Our great museums curry favor of the *uber*-rich and suck from them funds and concern that could serve more pressing social needs. A large, plastic replica of a shark in a pretend aquatic tank sold for $12 million. A skull painted with smeared blues and black sold for over $100 million. It is obscene. Any meaning that can be attributed to these "artworks" is overintellectualized nonsense that smacks of greed.

Where do we come from? What are we? Where are we going? These are the questions Paul Gauguin asked in 1898. Great art of the twenty-first century must reflect substance and ideals of humanism and the knowledge of science. As a matter of conscience, I decry the self-absorbed designs that pass for great art. I stand tall for the use of art to help us all understand our world and our universe, and to preserve it. I cry *foul*!

1

Vision for the Twenty-First Century

The Issue Is Art About What and Why

ART OF THE TWENTY-FIRST century explores the intersection of art, science, and ethics and is understood by its message to the viewer. The issue is art about what and why, and no longer about what materials are used, or the style of the artwork, or whether it meets some version of classical standards. Whether made with oil paints, acrylics, watercolors, flashing lights, electronic sounds, woven fabrics or materials, period costumes, street wear, frescos, canvas, urinals, jars of piss, tattoos, flags, golden toilets, or toy sharks, none of it matters except the work's message. Its style no longer counts, whether realistic, abstract, romantic, or street art with spray cans. What does it say? Does its message have substance? What is it about and why?

The twenty-first century's Scientific Revolution is here, now, and it foretells the accelerated and overwhelming impact of science upon our entire earthly world. The great artists of the twenty-first century will comprehend science as they seek the nexus of art, science, and ethics. As old ideas and obsolete paradigms of philosophical and religious thought are shattered, artists within their scientific knowledge might also birth a new sense of philosophy or perhaps religion that considers questions of life, death, and purpose that haunt the humankind. Can we trust our knowledge of the universe to guide us in finding a new sense of unity among all peoples and perhaps an inclusive, world philosophical or religious point of view?

The twenty-first century cries out for synthesis of scientific thought with the expression of ethics at its highest level, fostering a form of conceptual art at its ethical, philosophical best.

Within the universe as a whole, we do not yet know our true place or size. Those who cling to the illusion that humans have control over the universe engage in futility. It is a lost cause.

Science has brought enormous changes to our daily lives, beginning in the last half of the twentieth century. This change brought everything from computers, cell phones, and primitive space travel to the start of the World Wide Web. It has driven the world's population to learn new technology, with newborns grasping for electronics. The digital generation now rules. The Scientific Revolution brought not just some change, but massive, rapid change. With the twenty-first century came pictures from the Hubble Telescope that proclaimed the golden age of astronomy. Now we have even further evidence transmitted through the James Webb Space Telescope. We once believed we were part of the universe's one galaxy that we called the Milky Way. Now we have pictures of billions of galaxies. We have even pieced together a picture of the Milky Way from various angles.

In a salute to the foresight of the great artist Paul Gauguin, it is time artists ask, "Where do we come from? What are we? Where are we going?" Over 150 years ago, Gauguin had the courage to ask these substantive questions. Now, with the twenty-first century's Scientific Revolution, it is our turn to face these questions within the changing currents of scientific knowledge. The colonization of another planet is only a few generations away. Meanwhile, subatomic particles mimic each other within a ghostly, spooky dance of unknown energy.

The speed of change will accelerate. The world population has increased, and their ability to communicate has increased. This means scientific quandaries are considered more often by more people. It is a simple numbers game. The more people who consider scientific problems, the more thoughts and answers will be offered more often—faster and faster. As the speed of change accelerates, the ability of our institutions to deal with change needs to accelerate. Throughout the country, the older generation, not born to the digital lifestyle, clings to its power and rejects the young.

Unfortunately, this lesson seems lost on our philosophical and religious institutions just when a calming, ethical influence is most needed. As America becomes more secular, traditional sectors of the population such as rural areas of the United States—for example, areas found in Alabama, Oklahoma, Kentucky, Tennessee, Mississippi, Wyoming—reach back for "traditional values." They reach for perceived stability and decry change. They pray for easy answers. It is a lost cause.

The Scientific Revolution propels us into a great age of discernment, in direct challenge to artists.

From an ethical, philosophical, and/or religious standpoint (here expressly including art), we now enter the great age of discernment. It is a conundrum that with more scientific knowledge, we now have more questions. Artists are challenged to give our conscious minds images of where we come from, our true size, and where we are going. It is not just change that disturbs. It is the unknown. Suddenly, we have a much greater, personal, even intimate sense of what we do not know. In fact, humans know very little of the universe. We are so small that we have difficulty even finding ourselves within it. It evokes fear. Artists, aware of scientific discoveries, may assume the role of helping all cope with change and with the unknown, quelling fear so all may think clearly. So far within the twenty-first century, television and movie scripts have attempted these issues. Why do visual artists and art schools evade them?

Museums continue to collect and display beautifully colored designs with lavish, painterly brushstrokes. Popular among the so-called intelligentsia of art collectors and critiques are outrageous sculptures that imagine soft flowing lines from hard rock and fabric that squishes fun from fiber. But none of this will ever be great art. All avoid the obvious tough questions. It is like reaching for oblivion with a martini or a toke.

The designs of modern art that freed us from the shackles of realism are now dead. It is time for visual artists to participate in the Scientific Revolution as arbiters in the great age of discernment. It is a time to prepare humanity for transition to outer space.

Artists must seek a nexus of art, science, and ethics—and shatter the images and idols of traditional thought that now should be guided by our understanding of the universe.

Great art needs to be found within the realm of society's issues and not set aside within overintellectualized museums. It should be a catalyst that relates to all parts of every society, and at its best, all cultures. The purpose of great art—not just decorative art or fine craftsmanship—is to address the issues of our age. When viewing an image, the human mind can freely associate the image with all its stored memories. Free association is the autonomic response, unguided by text or speech, that humans use to relate what they see with prior experience. It is this very strength of art that allows the artist to pose conscious images of billions of galaxies across millions of light-years to a viewer's prior experience to help the viewer mentally assimilate it.

Presenting art that addresses society's issues confronts and substantiates that art for art's sake alone does not engender great art. While designs are helpful as studies to refine artistic techniques and to study how to paint, studies alone are not enough. In the end, they simply feed upon themselves as in the Dada art movement.

Much like a human maturing in mind and outlook, we all look within ourselves to understand how and why we feel and respond as we do, but the process of maturation does not stop there. The question is how we apply this understanding to the world at large. How do we get beyond the consideration of the self to a consideration of others? The artist should be a functional, contributing member of society, not some bohemian, self-indulgent "genius" that drinks and spews cheap philosophy and psychobabble.

Art needs to reestablish its social function. It should mirror current issues of where we are going and why. It should unite humanity; question ethics; give purpose. What are we? Where are we going? Why are we here? We will not have answers to these questions; only our attempts at forming them. For this reason, we truly live in the great age of discernment.

The central question is "Art about what and why?"

This is the core issue that great artists face. It drives those who aspire to great art and the issue that characterizes great art. With so much media available, we are drowning in information. Images are everywhere. The issue becomes what is important. What do we spend our time looking at and considering? As an example, if one took a photograph of everything one saw from birth until death, what makes some photos more important than others? How does anyone go through such a mass of information and select what is important? What is important and why? What is credible versus what is altered or positioned to a particular viewpoint? With the mass of information available in the digital age, the issue is, "Information about what and why?" In the art world, the question is, "Art about what and why?"

2

The Scientific Revolution Is Here Now, So Wake Up

The Scientific Revolution brings knowledge that births a New Matrix.

IN THE TWENTY-FIRST CENTURY, a New Matrix is born with the knowledge of our vast universe. This is its challenge: artists, art teachers, and art schools across all mediums including visual, written, and audio are called to address the questions of our age that really count! Eminent artists must synthesize scientific data to inspire others and to address issues facing all humanity and our planet.

Artists! Synthesize scientific data and present art that inspires coming generations who, like you, are born to the Scientific Revolution, and who must live with issues that have changed human perspective. Learn to incorporate science into artwork based on three criteria: (1) the synthesis of scientific principles, (2) artistic content, and (3) inspiring the viewer. Contemporary visual art that attempts to incorporate science generally uses data to suggest an issue while avoiding synthesis. This use of data resembles boring science fair projects. To synthesize and inspire calls artists to reach for the level of great art within the needs of the twenty-first century.

The Scientific Revolution has fundamentally transformed the global paradigm and brings a New Matrix of assumptions that shape how humanity understands itself and its size and place in the universe. This era of discovery has ushered in rapid and accelerating change, surpassing anything previously experienced in human history. Suddenly, we find ourselves immersed in a cosmos teeming with billions of galaxies, nebulas, and black holes, expanding the boundaries of our perception and redefining our sense of scale and significance.

A glance at a historical analysis.

It is hard to look down at your toes and know where you are. A compressed glance at the past emphasizes the enormity and speed of change within the twenty-first century. It may help awaken artists from their sleep and learn where they are.

Art is found in the earliest human-occupied caves. For example, consider the beautiful cave drawings of Lascaux, France, that exemplify human's earliest efforts of artistic expression about 17,000 years ago. Presented here in overgeneralized terms, human thought evolved as we entered the Stone Age with the use of stones for tools and family life in caves. Around 900 BCE, we developed into the Neolithic Age with its agrarian revolution and often a more stable, situated lifestyle such as Catal Hoyuk in Anatolia. Although grossly generalized, it is easy to sense that our large cities today are technologically advanced far more than the agrarian world of our early history. Glance at a visual representation of this time scale. Our present state where we locate our new Scientific Revolution is a mere dot found at the end of our timeline.

9000 BCE | 2025 CE

Although very short-lived, the Scientific Revolution holds strong impact with its rapid change and new knowledge that is of fundamental significance to all on earth.

The twenty-first century dawns the golden age of astronomy and space exploration. Few doubt that within a short time we will be landing on other planets. Our robotics are already there. Since the first moon landing in 1969—about .3 percent of the time since humans started agriculture—humans began space exploration. We talk, text, and watch TV on cell phones, use computers daily, Photoshop everything in sight, and use the web for everything from porn to college. Most recently, artificial intelligence brings computerized reason, research, and writing. All within the twenty-first century, yet this list omits the first real test of our humanity.

Human assumptions of our size and importance in the universe have changed forever. In the 1990s the Hubble Telescope began searching what we thought was a dark area of space—astronomers wanted to learn what was there, if anything. Hubble saw billions of galaxies, each with millions of stars, with many galaxies larger than our own Milky Way. Those unwilling to consciously embrace this knowledge are now obsolete, even primitive. Unfortunately, that includes nearly all the art world.

While humanity stands on the cusp of the Scientific Revolution, much of the art world draws designs. The changes that humans made from the Stone Age throughout the Agrarian and Industrial Revolutions will be small in comparison to the changes brought by space exploration. The past thousand years held few changes compared to those scientific discoveries of the last forty years. Think back to the years when Michelangelo painted the Catholic view of Christendom on the ceiling of the Sistine Chapel. The earth was assumed flat. Michelangelo portrayed God as an aging, gray-haired, Caucasian male. His work has stood as one of the greatest influences on Western art for five hundred years—but no more.

Envision for the twenty-first century a New Matrix within a nexus of art, science, and ethics.

Artists and scientists both seek a vision of the universe. Every aspect of all that we experience and know is open to their inquiry. Within the twenty-first century great artists will integrate both science and art. To some degree artists will function in both realms and seek the nexus between art, science, and ethics. The universe, by its inclusion of all peoples and creatures, calls for an inclusive view of ethics not bound by one point of view, not tribal, not cultural, but rather an expression of ethics at its highest level of global, planetary, and universal concern. In this century, those whose spiritual direction seeks to discern a Creator through the arts may well consider their commitment to a higher power within a religious covenant born from this New Matrix of art, science, and ethics. As Abraham shattered the idols and images of his time and birthed the Jewish tribes, so our new knowledge of the universe calls us to shatter our outmoded paradigms as we contemplate ourselves among billions of galaxies and our exploration of the stars. Artists must help with this journey.

The next generation of great artists will understand science and to some degree be scientists themselves. Well ahead of his time, in *The Sciences and the Arts*,[1] Cassidy argued that scientific thought can be reduced to three general forms:

1. *analytic* activity that involves accumulating data;
2. *synthesis* of that data that seeks connections and formulates explanatory theories; and
3. *reduction* of these processes to practice or the activity which directs thought from the general to practical application.

It is two of these thought processes, the analytic and synthesis, that distinguish art and science. While both art and science engage in all three forms of thought, art is able to focus more on presenting synthesized thought with its sense of inspiration.

1. Cassidy, *Sciences and the Arts*, 21–23.

Science focuses more on analytical thought with engineering and other areas focusing more on reduction or practical application.

Current integration of artistic and scientific thought tend to fail at the level of synthesis. While attempting to show scientific data or facts with interesting displays, such efforts lapse into the realm of mimicking science fair projects and illustrating facts with concrete thinking. Art needs to synthesize science and address larger human issues. It is the level of synthesis that inspires.

To synthesize data means to give expression to complex ideas and to express the relationship between ideas in a simple and direct manner. This requires that the artist be an integrated person who thinks both in the logic of science while also able to interpret and synthesize ideas with feeling and the compassion to inspire. Artists do not need to do science but rather use it to give their works substantive meaning.

To inspire others emanates from a developed sense of ethics. The essential questions humans face are now defined by both science and ethics—or if you prefer, religion. Who am I? How did I get here? What is my purpose? None of these issues can now be faced without including scientific knowledge of outer space, thousands of galaxies, and a true estimate of our size—less than a speck of sand in the universe. To some, this includes recognition of a Creator or an unknown energy source.

The twenty-first century as seen through the lens of science provides a new context for developing artistic expression through science and ethics. The awareness of earth as our planetary spaceship that circles within a universe we have no control over is the basis for artistic inspiration and contemplation. It is the commonality for all, no matter what religion or non-religion or culture, who search for the ultimate purpose of life's experience. It is the basis for great art in the twenty-first century.

Those substantive issues that an artwork evokes provides the basis for evaluating and valuing an artwork. Within scientific experiments, it is analogous to a baseline. It confronts the issue of what the purpose and meaning of a particular artwork is. To evaluate and compare artworks, there must be some means of measurement. Rather than avoid ethical thinkers, artists need to embrace them and make ethics their own journey by which they can be seen and evaluated. Embracing the need to evaluate what constitutes artistic value does not suggest stereotypes or prejudicial thinking, but rather it suggests honoring those who consider universal and ethical questions posed to develop human consciousness. The twenty-first century calls artists to contemplate the place of humans within the universe, similar to Gauguin: "Where do we come from? What are we? Where are we going?" And now we add, "What is our purpose? What is our true size? Who are we? Is there a Creator or an unknown source of energy?"

Artists might consider pictures of the Hubble and James Webb Telescopes that show billions of galaxies and the birth and demise of stars. These pictures provoke and support thoughtful humans to ponder their true size. How can we accept ourselves as a minute part of the universe that is vast, humbling, and unimaginable? Can artists present this experience visually? Can artists help people consider whether our being is such a part of the universe that our consciousness, however small, provides a record of what happens here, of its beauty, energy, and development, including our development of ethics? Is the energy of the universe, as it generates new stars and collapses old ones, the transformation of energy from life to death and life again, as posed by some religions? Is death really transformation, most immediately of ourselves but also of earth, our galaxy, and the entire universe? If the universe links human destiny, are artists required to imagine the universe so others can internalize their own understanding of it? How do artists help in this process, especially if they are asked to communicate images of the universe to those who speak another language, or do not read or write, or do not know anything of vast outer space?

Great art will synthesize questions that count.

Art must reach to synthesize scientific data to inspire others. Art is not an imitation of science. In fact, science can be destructive to art if considered only at the level of analytic data. Art is not science fair projects. It is the visual expression of values, quandaries, and conundrums that face each person that walks the earth—or inhabits the universe. *This is the challenge. Artists, art teachers, and art schools: address the questions that count.*

3

Scientists and Artists Are Both Visionaries

And Both Must Heed the Role of Ethics

ALBERT EINSTEIN WROTE THAT the mystery of the universe "stands at the cradle of true art and true science." His life's pursuit of knowledge was fueled by a profound engagement with the mysterious and unknowable beauty of the universe. This pursuit now stands as the foundation for the twenty-first century and its Scientific Revolution. *Read this quote that conceptualizes Einstein's nexus of art, science, and ethics/religion.*

> The fairest thing we can experience is the mysterious. It is the fundamental emotion which stands at the cradle of true art and true science. He who knows it not and can no longer wonder, no longer feel amazement, is as good as dead, a snuffed-out candle. It was the experience of mystery—even if mixed with fear—that engendered religion. A knowledge of the existence of something we cannot penetrate, of the manifestations of the profoundest reason and the most radiant beauty, which are only accessible to our reason in their most elementary forms—it is this knowledge and this emotion that constitute the

> truly religious attitude; in this sense, and in this alone, I am a deeply religious man.[1]

Throughout time all humans have asked fundamental questions with inherently unknown answers. What existed before time? What or who made or generated the known universe? Why do we die? What happens to our consciousness after death? Do we have souls that live beyond death? Among the earliest human drawings and paintings created around 17,000 years ago in the caves of Lascaux, France, the running bulls and antelopes still resound with their quest for the unknown. The ancient wisdom of the Bible speaks to the same yearnings. In the Gospel of John at chapter 17, verse 24, Jesus is quoted over two thousand years ago, "Glorify me . . . with the glory that I had in your presence before the world existed." What was there before the world existed?

In the twenty-first century, our quest has fundamentally changed. Both scientists and artists face humanity's new engagement with the mysterious and unknowable beauty of the universe "at the cradle of true art and true science." Both must experience these questions and use the knowledge and imagination of their discipline. Both share in a profound passion to understand our known universe and share in the grit to learn.

Without ethics, both science and art are meaningless.

Science without ethics can be depraved.

Science provides a methodology for learning that addresses how physical laws of the universe work. Science does not provide nor pretend an internal compass or ethics of any kind. That is the job of the human conscience. One can mathematically postulate $e = mc^2$ but not how to use the atom bomb. Medicine saves lives but also gives us poisonous gases. Machinery makes tools for farming and also bullets and guns. More poignantly, it was science that built the

1. Einstein, *World As I See It*, 16.

ovens of WWII genocide—depraved scientifically, and ugly and meaningless artistically. Enough said. Einstein himself sought an ethical basis for his scientific thinking: "Try not to become a man of success but rather try to become a man of value."[2]

Science alone is not enough and few claim it is. Many scientists are deeply involved in the search for values. For example, the Cambridge Roundtable on Science, Art, and Religion struggles with questions such as how we can use the ethical basis for religion to form our values for exploring the universe. All humans are imperfect. We strive toward our own development, yet human failing permeates all. However, note that it permeates *all* theories and *all* efforts at ethics, not only the scientific. At least science requires observation and data and the ability to reproduce similar results. While science seeks to understand physical laws of the universe, humans provide the conscience of self-reflection—the ethical basis for the use of scientific knowledge. At its best, science should describe the world as it is in reality. It is artists within a developed sense of ethics who can offer the use of imagination for the good of all, create beauty, use color, and envision something new for the betterment of all.

Science gives us humility as our compass.

Both scientists and artists are called to discern with clarity what they know from what they do not know. It lies among the greatest acts of humility. We now can imagine our true size within the vastness of the universe, and all are called to confront it. Compared to the universe, humans are minute. When we survey our beloved Milky Way galaxy in the night sky with its millions of stars, the size and scale of the known universe is staggering. Our human mind has difficulty integrating it into our limited view of life on earth, and we balk at its concrete conception. It is humbling.

Any thought that humans control the universe is misplaced. An understanding of the subatomic world in which one particle

2. "Interview with Albert Einstein," *Life Magazine*, May 2, 1955.

mirrors the movements of another at a distance without apparent communication remains beyond our understanding. Yet it is certain, humans are not the center of the universe. Humility itself must be our compass, and its conception must be within the role of both artist and scientist.

> Artists are called to symbolic expression of
> self-reflection and substantive ethics.

The personal development of an artist as thoughtful and thought-provoking is exposed for all to see in their artworks. Without development, without thought, without basis for belief, artwork reflects the vacuous lack of thought and underdevelopment of the artist. As humans mature, the reality of their mortality seeps into their lives, and most hear the march of days as they pass. How one addresses this issue forms the basis of a mature adult. It is through the development of character sounding in ethics that first we learn to think of others rather than only the self. This enables us to learn understanding of others and form compassion for the world at large. So often in life, personal grief, struggles, or illness ignite this period of growth and revelation. It seems the human condition creates overconfidence and conceit from having a gifted life handed to us, for it is struggle that often motivates development. Great art is meant to reflect substantive thought and clear vision. Einstein explained that he conceived detailed visions and images of his theories first. Then he worked the math. Remember his detailed example of traveling by train and how he understood its movement in relation to the land. In whatever manner one develops and reaches to find deeper thought and meaning, artists must take this journey.

Both artists and scientists are called to envision a future through a disciplined inner life that contemplates the mysteries of the universe and humanity's relationship with it.

While respecting the knowledge that science reveals, much is unknown, and the power that our inner lives bring to the universe remains unmeasured and mysterious—except for those who experience it. There is no scientific explanation for the power that a forged inner life of disciplined thought and meditation brings. We do not know it in scientific terms—yet. Perhaps one day we will measure it as we now measure X-rays and gravitational waves. The untapped resources of the universe were apparent to Jesus, Buddha, Allah, Moses, Einstein, Martin Luther King, Gandhi, Julian of Norwich, Jose Silva, and many others, yet somehow remain unproven and even unaccepted in the scientific world. Philosophical, ethical, and religious thought tell of humans who experience energy or perhaps an unknown presence within the universe, and surmise miracles within moments of daily life. Those who have experienced this express this phenomenon as real as science itself.

Science affirms that humans are not separate beings on earth or in the universe, but an integral part of it. We exist within the system or relationships of the universe. While we experience life on earth as separate from it, in reality we are a part of it. Humans function as all other energy, changing and altering in the universe according to its physical laws. All our actions have an impact upon the universe, if even minute. However, unlike all other energy (at least that discovered so far), human beings have the ability to be self-reflective. Perhaps we may yet save ourselves.

All humans experience life in limited form. For example, we cannot see X-rays or radio waves or gravitational waves, but we can measure them and know of their impact. Similarly, the internal life of those who pray or meditate or who simply experience telepathic thought or energy is real. Its results are felt. Many attest to the power of disciplined thought and meditation. Einstein thought in visions and then translated his visions into mathematics. The scientific method alone, at least so far, does not tell us everything.

It is a fundamental conundrum that while truth lies within the soul, it surely needs the foundation of scientific thought and knowledge. Often, this conundrum leads to an argument for atheism; however, in contradiction, it may also point to our creation as sacred and foster scientific and artistic thought that envisions ethics grounded in a united planet and reflects its relationship to the universe. Artists grounded in a scientific understanding of our universe are called to self-reflection and substantive ethics given expression through art.

Art and science share spiritual and ethical insight.

While the benefits of science often lead to arguments on behalf of atheism, in contradiction, it may lead to the more humbling acknowledgment that we do not know much, and that what we do know of this life is wondrous and amazing, if not outright sacred. It is difficult to sustain a discussion about ethics symbolized in art without considering spirituality. But cloaking our amazement in superstition and sophistry is no longer agreeable as we seek our destiny within the Scientific Revolution. Our ethical values should help us cope with the world as it really is, not how it may be imaged within a guise of philosophical sophistry.

Art has long been influenced by various religions, and all major religions have given artistic expression to their beliefs. The Egyptians built pyramids to embrace their dead; the Greeks built temples to their various gods with world-renown sculptures; the Buddhists built massive images of Buddha; the Jews built the Temple of Solomon; Muslims have ornate mosques; the Vatican holds some of the Western world's greatest artworks; and Christian cathedrals abound. The realm of the spiritual and religious have deeply enriched art across many cultures over centuries, and the arts themselves have offered viewers connection to heartfelt feelings of spirituality without the demand of religious law or theology.

Within the twenty-first century, art may now draw upon current scientific knowledge of our universe to unite all people. We exist in one world within one reality. As artists search for the nexus of art, science, and ethics, they will grapple with the impact of religion on ethics and the need for a broad vision that confronts the vastness of the universe beyond mere earthly concerns. Artists may confront whether the serious development of ethics can be engaged and sustained outside of religious concepts. At their best the major religions of the world provide a thoughtful construct for this experience. Unfortunately, not a single major religion has stepped up to the challenge of incorporating our current scientific understanding of the universe in the development of their religious commitment. Can we draw our understanding of a Creator or that of a creative energy from our knowledge of the universe? Does the universe itself offer lessons in ethics or even an understanding of death and renewed life through death? The universe holds both profound knowledge and profound mystery. Einstein wrote that this knowledge and emotion for him formed the truly religious attitude. It is the mysteries of the universe that call to the vision of both artists and scientists. It is the nexus of art, science, and ethics.

It is this nexus of art, science, and ethics that poses the great conundrum of the twenty-first century. While science grounds our understanding of the known universe, our instincts continue to lead us, and our truth is heard by the soul.

Many ethical bases arise from spiritual precepts. Most of us have heard some version of this before: "Love your neighbor as yourself." It exists in all major religions. Within this golden age of astronomy, we are unified by the visible knowledge that we have only one planet. We must adopt a global identity beyond any quest for power. We must find reconciliation and reformulate our values to serve the planet. Either we all win, or we all lose.

Planetary thinking in ethical terms grounds artists and scientists for considering issues of climate change, pollution, food, and overpopulation. For religious and political reasons, few speak of global overpopulation and how it relates to the energy crisis, health issues, water crisis, housing, adequate food, and on and on.

Where are the artists? If artists are really the so-called rebels, as they so love to color themselves, why are they not speaking out on the "elephant" that permeates all political and religious discourse? At what point are self-regulated civilizations willing and able to aid those societies that are insistent on uncontrolled population growth without the ability to support themselves? Is there any art in this country that considers these questions? Not much. Fully abstract art creates pretty designs and lovely colors, sometimes conveying the opposite effect of extreme outrage. The great art of this twenty-first century must look to substantive issues and challenge both artist and viewer.

In the year 1250 CE Thomas Aquinas asserted that both reason and revelation together forged truth. Still today, scientific truth is learned through reason, yet the Spirit, by whatever name, is still heard by the soul. Startling! At first glance, this contradicts the basis of science, but the inherent amazement of life itself as expressed throughout time tells of the mysterious and soulful essence found at the nexus of art, science, and ethics. *The Spirit of Truth and our discovery of the universe's mysteries lie at the nexus of art, science, and ethics.*

4

An Offer of Inspiration

NASA Images

NASA OFFERS PHOTOGRAPHS TO the public from the Hubble and James Webb Space Telescopes that are suggested here as inspiration for the unfolding of twenty-first-century compelling science. Generally, all NASA photos of outer space are within the public domain and may be copied to inspire those who wish to consider their implications. Although the NASA logo and pictures of its employees and astronauts may *not* be copied, for outer space photos, NASA has a sophisticated website called NASA Image and Video Library that is inspirational in combining levels of art, science, and ethics.[1] Paraphrased below are key points from NASA's media usage policy:

1. NASA content (images, videos, audio, etc.) is generally not copyrighted and may be used for educational or informational purposes without needing explicit permissions.
2. NASA insignia logo (the blue "meatball" insignia), the retired NASA logotype (the red "worm" logo) and

1. NASA, "NASA Image and Video Library."

the The NASA acronym may not be used for any purpose without explicit permission.
3. When using NASA content, try to use NASA photos.
4. Give attribution to the required NASA website (they have over 20 public-facing sites).
5. Use only the amount of audio or video that you actually need to illustrate your point.[2]

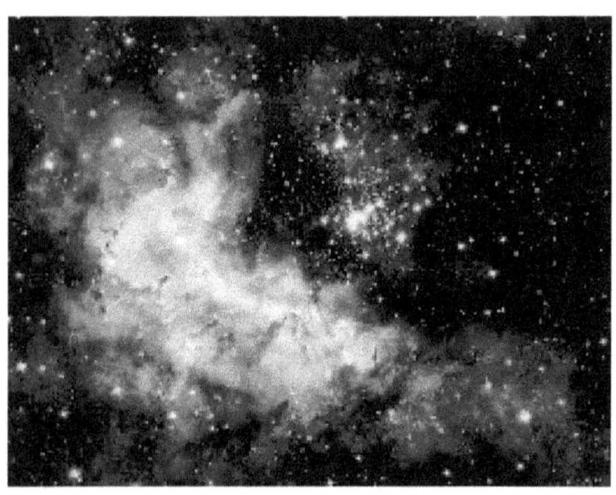

Witness the powerful birthing of stars.
It is also the birthing of beautiful abstract designs sounding in science.
Find many, many more:
Search: "NASA photos of space"—or
Link: https://stock.adobe.com › photos

2. NASA, "Images and Media."

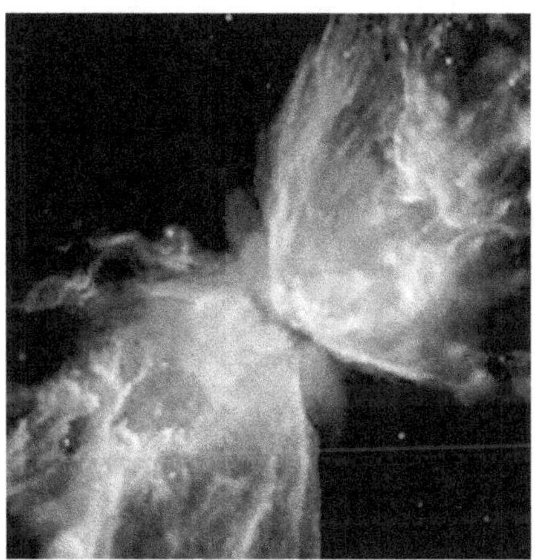

Try playing "Name that Nebula" on the NASA website.
This one is the Butterfly Nebula. Fun.
Search: "NASA photos of space"—or
Link: https://stock.adobe.com › photos

5

A Plea for Artistic Content and Thought

Art must move beyond self-absorption.

WITH MATURITY, SELF-AWARENESS EXPANDS to an understanding of others and an awareness of the world beyond the self. Along with the advent of self-help books, psychoanalysis, and generally minding everybody else's business, artists creating designs from the subconscious now dominate much of the art world. Without question, understanding the self is a valued expression. Art therapy, or even a child's drawing, is an experiential wonder and fills valid need. However, for the professional art world to stop there is not enough. We all explore the self, both children and adults, and rightfully so; however, it is a problem when the self is all that is explored.

Consider these questions as you ponder even one photograph from an outer space telescope: Who are we? What is our role in the universe? What is the universe? We can see the universe give birth to stars and planetary systems, and the death of stars as they are consumed by black holes. Does this imply that the universe has its own consciousness? Is our purpose found as a self-conscious, discerning element of the universe, despite our poor job of it? We know that space has all the needed ingredients for carbon-based life. Are we the only form of life in outer space? How big is the universe and where are its boundaries, if any? What could exist beyond its borders? Is our consciousness the only plane of existence? On our own earth, artists can contemplate whether peace is of greater benefit than war. Is war ever justified? See Picasso's *Guernica* as an example of visual, substantive thought. Are we burning ourselves up as the climate warms? Abstract artworks of design are self-absorbed without ethical basis other than self-awareness or studies of color and line.

As adults, we are ethically bound to respect others, to love our neighbors as ourselves, to get along with others (my thanks to Rodney King), and to seek understanding of others. I fully support being self-aware; however, self-awareness does not stand alone. Like all children, including artists, our developmental journey starts with expressing the self. As mature adults, at our best, we grow beyond the concept of self and begin the journey of understanding others. It is ancient wisdom. While a child, I played with toys and childish things. When I became an adult, I set them aside.[1] It is time for artists to conceive of themselves as mature adults and set aside childish things. It is time to address meaningful issues. Art should move beyond self-absorption. Great art must.

The historical approach to art analysis fails to confront rapid scientific change and the need for ethical response.

Most art history analysis continues in a traditional, historical approach. Some analyze only artworks. Some analyze artworks within historical context. Some expand this analysis—as if it were a different approach—to include social and political influences. It may be a path for understanding the past, but with the advent of

1. 1 Cor 13:11 (NRSV).

the Scientific Revolution, it all fails to hold a vision or direction for the future of art.

For example, Shiner's *The Invention of Art: A Cultural History* is an excellent historical analysis of art and places it within context, yet by design it exemplifies excellent analysis of the past with no vision of the future.[2] Like most current analyses based in historical context, it literally points us nowhere. It does not suggest current trends, a hint of direction, nor ask a single question of where we are going in art or any other context.

Traditional art history analysis does not consider two main factors. First, it overlooks that current culture is driven by the gigantic scientific leaps that impacted the beginning of the twenty-first century. With human perspective of the earth and the universe now fundamentally changed, the twenty-first century fundamentally differs from the past. This change is beyond a philosophical difference. It touches the daily activities of people worldwide through cell phones, computers, streamed entertainment, web information, AI, medical advances, microwaves, and much more. Even our toilets are now self-cleaning. Take that, Marcel Duchamp! Your urinal is totally outdated. The impact of science on the world is overlooked by the traditional art historical analysis approach. The educated elite have begun space exploration. Where are the artists?

The second factor that is missed by traditional art history analysis is its lack of ethical response to this fundamental change. Contemporary ethical thought struggles to respond to scientific knowledge and find relevance. Art reflects this. No one is asking the tough questions. Why did you draw this? What is your point? What does this mean? What inspiration does it provide? What is it analyzing?

Change is now so dramatic and so rapid that most historical analysis of art—and, in truth, historical analysis of virtually anything—fails. There is a new paradigm afoot, much like a new sheriff in town, and it is *science*. What is needed is a vision of the future. Art should be judged on its contribution to finding this

2. Shiner, *Invention of Art*.

vision. Museums in particular need to heed this problem. French impressionism plays only to those who are interested in history or perhaps find themselves obsolete, not engaging in the digital lifestyle and not embracing universal questions. It is old. Donors may be drawn from this demographic (as discussed in chapter 10), but the future will not be.

Examples of "environmental art" demonstrate a failure to find the nexus between art, science, and ethics.

Most so-called "environmental art" works generally depict science fair projects. The now-famous *Leonardo* journal, published through one of the finest scientific institutions in the world, the Massachusetts Institute of Technology (MIT), advertises itself on its website as "the leading international peer-reviewed journal on the use of contemporary science and technology in the arts and music and the application and influence of the arts and humanities on science and technology."[3] In truth, it illustrates why the future of art as it seeks to enrich our lives should *not* be left only to the scientists. So boring. It generally approaches art on the level of data rather than synthesis. With all due respect to scientists, stay in your lane; artists, step up! The integration of artistic sensibility and scientific knowledge is not fostered by this penchant for presenting data as if it were art. It is not enough. Great scientific and artistic thought merge in a well-considered vision of the future. Both artists and scientists are visionaries, and together must seek the nexus between art, science, and ethics.

Consider the example of so-called "environmental art." Overall, most of what passes for environmental art is limited in scope and generally offers data rather than synthesis. Few synthesize scientific thought and offer inspiration while addressing a significant environmental issue. Some excellent efforts are made to remind us

3. Massachusettes Institute of Technology, *MIT Leonardo Journal*, https://www.leonardo.info/.

of animals and plants nearing extinction or now sadly extinct. For example, Todd McGrain's 5 sculpted extinct birds are displayed in the Smithsonian Museum. Similar efforts tend to present realistic images of what is lost, rather than a synthesis of ideas for the future.

As we now face the most severe environmental crisis in human history, including the energy crisis, environmental warming, overpopulation, carbon dioxide pollution and the like, something is fundamentally wrong. Substantive content in art has been minimized. Amateur imitations of science fair projects do not meet artistic or scientific standards. I suggest that most artists fashioning environmental art works have limited understanding of the scientific basis for them.

For example, Wim Delvoye's pretend poop-maker *Cloaca* is not science. It is damn funny with potty humor to match that of any third grader. (OK, I laughed at it myself, but it is not great art.) Delvoye seemingly fashioned his poop machine from whimsy—or the obvious line that he pulled it out of his rump. At one end is input; the other out-put; and in between fanciful gismos of grinding and squishing. Imagine the Sistine Chapel with his poop machine on the ceiling or perhaps standing in a piazza in Florence, Italy. Funny, but is it great art, and where do we go from here?

"Environmental art" essentially means "environmental protection art."

The Library of Congress figured it out, and that is how they now classify art about environmental concerns. The term "environmental art" is just too broad. The "environment" covers nearly everything on the face of the earth. Environmental protection art integrates the sciences and the arts. I suggest that artists use scientific subcategories to reference environmental art, i.e.: "environmental art re: climatology." This may help artists to focus their efforts and integrate science and art with politics, ethics, economics, and the like.

Initially, environmental art seems to have begun with the land artists of the 1970s–90s. From this it seems to have devolved into quasi-science with less and less political messaging. For example, those works found in Beardsley's *Earthworks and Beyond*[4] such as the *Spiral Jetty* and as found in Ede's *Art and Science*[5] confirm that artists have generally engaged in macabre science fiction, as if science fair projects gone awry.

Some artists have chosen to use the designs found in scientific observation to invent imagined designs of great beauty. It approaches a new form of realism such as Susan Aldworth's *Brainscape*[6] and Andrew Carnie's *Magic Forest* of neuron patterns that he colored to look like trees.[7] While beautiful, these offer no vision of the future and not any real understanding of science. They both fail on both fronts. They are studies that may pave the way for great art, but not great art in and of themselves.

Artists and scientists are both visionaries, and artists and art schools need the focus of substantive issues.

Artists and scientists are both visionaries, yet artists have seemingly lost this. Instead of thought, they substitute wild imagination, as if creativity and originality are a substitute for disciplined thought. Leonardo drew a helicopter when it was still just science fiction, yet it was disciplined thought. I suggest that artists need to return to disciplined thought and present a vision within twenty-first-century issues. Presenting quasi-machinery or use of light is interesting, but it does not suggest a vision of the future.

To make art that synthesizes scientific thought, the artist must be trained in science and have a scientific understanding of the basis for the art. This leads to a fundamental change in how

4. Beardsley, *Earthworks and Beyond.*
5. Ede, *Art and Science.*
6. Wilson, *Art and Science*, 68.
7. Wilson, *Art and Science*, 68.

art is taught, and how artists are educated. Since some artistic minds—those of us less than the genius of Leonardo—have difficulty embracing the ridged scientific thought process, some means needs to be at hand for artists to train in science. Art schools are not prepared to offer any help in scientific training, and so the scientific world needs to allow young artists to train in science without facing a potential academic disaster—like flunking out of college. As discussed in the next chapter, there is a fundamental problem between the education that artists need in both science and substantive design and the education provided in art schools.

6

Artists and Scientists Need Each Other

Both scientists and artists are visionaries, and both need each other. Together, they may bring a vision of the future, synthesize ideas, and inspire each other. Does the brain function differently when engaging the artistic process as opposed to the scientific process? It seems likely that it does unless one has the genius of Leonardo himself.

The famous MIT *Leonardo* journal[1] is inundated with science projects that aspire to pass as art but are ultimately flat, boring, and unattractive. While providing good training to push students' thinking towards more creativity, it does not foster great art, and likely is not intended to do so. Caltech had an art department that suffered the same fate. Today, Caltech can share its scientific knowledge with the Pasadena Design Institute, a respected art college. Why do engineers and scientists seem so boring to artists? And to scientists, why do artists seem so overwrought with

1. Massachusettes Institute of Technology, *MIT Leonardo Journal*, https://www.leonardo.info/.

feelings and intuition? These questions bring us to the crux of the issue. They need each other.

Consider the example of NASA, the National Aeronautics and Space Administration, which presents an elaborate collaboration of art and science. Telescopic photos of outer space tell all humans of their minute size in comparison to the known universe. These photos are astonishing. How is this communicated to those who are not involved in astronomy, yet whose daily lives are formed by the universe and its impact upon the earth? How do individuals integrate this information into their psyches? How big is the universe, and how many of them are there? NASA has developed a massive collection of material brought from outer space that is available to the public and often colorized by artists to teach the lessons now confronted by the twenty-first century. Yet throughout all this, the question of ethics seems to hang in the abyss, unmentioned. The United States Space Force now operates within the US Air Force "to secure our nation's interests in, from, and to space."[2] NASA and the Space Force have a long history of cooperation.[3] Where are the artists in considering the ethical consequences of all this? They are unknown.

Artistically, NASA has compiled a vast amount of information with artists not participating in synthesizing issues, communicating them, and confronting their ethical challenges through artworks. It is a process that artists have not engaged while focusing on abstract designs. NASA brings only one set of issues that face humans on earth. This one example alone calls artists to excellence alongside that of scientists.

2. United States Space Force, "Space Force Announces."

3. For example, search NASA's website at https://www.nasa.gov under their many topics for videos and images. NASA maintains a large image and video library open to the public for online access.

7

Juvenile Narcissism as Art

As argued in the next chapter, art for art's sake is dead. Pure abstraction has passed for great art in the United States, yet it amounts to little more than juvenile narcissism done in lovely colors. Wholly abstract art form makes no sense. Alas, even the US Postal Service has honored this purely abstract art form with stamps touting supposedly great American artists. I cry foul to the entire genre.

Such works are not great art because they have no demonstrable meaning, nor do such works move beyond the artists' sense of self. While these works may artistically represent studies in color and perhaps personal growth, they do not rise to the level of representing our nation, nor are they great artists. I decline to include their works to avoid any further effort to artificially inflate their prices or enrich those who trade in them. See their images online. Every artist enjoys a good scribble, but do not pretend it is great art.

Mark Rothko (with reverence)
Hans Hofmann
Joan Mitchell

Arshile Gorky
Jackson Pollock
Adolph Gottlieb
Robert Motherwell
Clyfford Still
Willem de Kooning
Barnett Newman
Find many others at your leisure.

8

Art for Art's Sake Is Dead

Substance Lives On

Art's Role in the Virtuous Circle

ART AS AN AUTONOMOUS enterprise has failed. It cannot sustain itself external to society as a playground for the indulged inventive. Art should uphold a substantial role integrated within society both national and global, commenting upon it, addressing issues, and enriching lives. It should explore our values, speak to our lives, and struggle alongside us all. It should enlighten people, improve society, and reach for global impact, all within a consciousness consistent with our scientific understanding of our earth and our universe. This is a huge, idealized goal! Former President Bill Clinton's Conference on Global Initiatives tackled such issues every year: development of women's roles, sustainability of resources, and the like. In an article for *Time* magazine, October 2012, Clinton said, "I firmly believe that progress changes consciousness, and when you change people's consciousness, then their awareness of what is possible changes as well—a virtuous circle."[1]

1. Clinton, "Case for Optimism," 38.

Freeform design is by definition arbitrary.

Anyone can imagine. Anyone can create a design. Within pure abstraction, this is its strength and its weakness. If anyone can do it and it is by definition freeform, what standards does one use to evaluate it other than "I like it" or "It fits with my décor"? There are no standards for evaluation. There is no way of measuring it or determining which freeform is better than another. Why does a line travel in one direction and not the other by any explanation other than random thought, subconscious motivation, or some neural connection within the brain developed who knows when for whatever reason? When creating a design, when are you done? What constitutes a conclusion to the design other than emotional exhaustion? Why is one design better than any other?

I have never forgotten an exercise for postgraduate art students at the School of the Museum of Fine Arts in Boston. By random drawing, three student artists were chosen to display their artworks to two professional art critics. All three art students worked in the ever-popular realm of abstraction. The art critics were isolated from each other and stated their views outside of the hearing of each other. As I listened to their comments, not one single word was even similar. Despite their qualifications as professional artists and art critics, each had disparate views of the artworks. Freeform design is perceived through free association of color and images of both artist and viewer and lacks any standard for comparison. Other than execution of the media used, there is no means to evaluate purely abstract art. Evaluation requires substantive content.

Pure abstraction equates to pure artistic expression.

This is the one value of pure abstraction. If eyes are the window to the soul, pure abstraction is its visual expression. It is the pure inner expression of the artist. If one chooses to engage the psyche of the artist, wholly abstract art that visualizes thoughts, feelings, and intuitions of the inner, subliminal self are important. By the mid-1900s, after art's emphasis on realism, romanticism, and the demands of French academics, artists reached to be free of constraint. Pure abstraction came as a contrasting relief.

To the artist focused on technique, pure abstraction can generate new ideas such as use of color, paint, or scale, much like brainstorming an idea. However, it is an expression of self rather than substance beyond the self. Perhaps pleasing for the artist or viewer, pure abstraction does not qualify as great art.

For example, consider the dark canvasses of Mark Rothko's Chapel that in reality portrayed his severe depression. The art world touted these works as great art, when in fact they were a cry for help that the overintellectualized art community missed or ignored. Rothko killed himself in 1970, two weeks before the canvases were placed on public view. These canvases did not portray wonderful artistic endeavor as much as severe mental illness. Perhaps this was an attempt to help himself and thus validate art as a therapeutic effort. It does not validate these canvases as great art.

Even beyond therapy, abstract art is fun. Smearing paints around is fun. Having a laugh over a design is wonderful. Enriching one's home with color or design may offer comfort. But is this great art? Is it art that belongs in our national museums and evokes feelings of the human soul? No. We live in the age of the Scientific Revolution and the golden age of astronomy. Pure abstraction has lost its value. Abstraction is the expression of the self. Humans now face questions beyond the self.

Global issues and the artist

Global issues are taught in high schools and undergraduate colleges across the country. An example of such education is the concise, well-written teaching aid entitled *Educating Globally Competent Citizens: A Tool Kit for Teaching Seven Revolutions.*[2] In fact, one of the editors was formerly chairperson of the art department in a rising four-year college, Fort Lewis College, Durango, Colorado.

2. Falk et al., *Educating Globally Competent Citizens.*

The issues were clearly set out for students to consider, but the art students at the college were not required to address them in art class. The art students continued to use wire to shape shoes and hallways to learn perspective.

The challenge is to teach artistic skills along with the ability to address substantive content that matters. This "tool kit" (as it calls itself) points to seven issues: population, resource management, technology, information, global economic integration, conflict, and governance.[3] Teaching tools are provided. So why not use them to teach art? Why not challenge art students to participate in global issues? Why not require art students to participate in the Scientific Revolution by learning basic science and collaborating with scientists to help us all build a new conscious image of what we are, what size we are, and where we are going?

Collaboration required.

The integration of art with science and ethics cries out for the collaboration of people with different backgrounds. Michelangelo worked with priests that determined the theology he painted. He did not work alone, and there were many that helped. No great artist needs to work alone, but use caution. Choose your collaborators well. Choose a scientist. Choose an ethicist. Choose a priest/monk/nun/therapist, but choose someone with mature substance.

Great art calls to virtue and to science.

Picasso's *Guernica* famously combines both abstraction and content. It stands as an emotional response to the ravages of war that exemplifies great art. Now artists must step up and help all to conceive of human purpose visually and emotionally from the newly discovered depths of the oceans to the galaxies.

Science drives the twenty-first century. It propounds questions at all levels, from the quantum to the galaxies. Michelangelo

3. Falk et al., *Educating Globally Competent Citizens*, 5–6.

painted when people believed the world was flat and the sun revolved around the earth. How do humans conceive of their minute existence within a universe so vast as to be incomprehensible? Is the universe alive? Does the universe present and demonstrate the ethics central to life itself or perhaps the ethics of its Creator? How does one internalize an understanding that the next galaxy is millions of light-years away? How can we save our earth from the heat of climate change? How do humans confront these issues with common values that sustain life and our planet? Why do artists, art schools, and museums sleep when the greatest issues humans have ever faced confront us?

9

Implications for Art Schools

The Search for Substance

WITHIN THIS CENTURY'S SCIENTIFIC revolution, artists need to understand, respect, and visualize science. It is the search for substance. In the past, great artists painted the world as they knew and understood it. Now we must paint it as we know it. Art schools and colleges need to meet the challenge of teaching students a basis for scientific substantive thought. For example, art schools could offer interdisciplinary education including coursework in the sciences such as degrees combining studio art with biological sciences, environmental sciences, engineering, and astrophysics. Where would a book such as the very one you are reading fit into this structure? Academia already has available course expertise but needs a more flexible degree structure.

Art schools could encourage artists to engage science without the threat of academic failure. Many artists are not mathematically or scientifically inclined, and these areas are hard for them to learn. Few have knowledge of physics or biology, or engineering and the like. They understand color more than calculus, yet they need training. Can mathematics, astronomy, and physics professors expose artists in general terms to their disciplines? Can art students receive class credit for watching DVDs or videos that explain science and then use this knowledge in their artwork? For example, allow an art student to take a science course, paint visualized concepts that inspired the artist, and receive class credit.

Few art instructors are trained to teach anything about science, and some snub it as beneath them. However, it is the integration of both scientific and artistic vision that calls to art for the twenty-first century. Both art teachers and art students need training in areas of science.

There is such a gap between art and science today that the Massachusetts Institute of Technology (MIT) has begun to train scientists in art, but not the other way around of training artists as scientists. This great institution only offers art to undergraduate scientists. They do not have a degree in art at any level. MIT even trains its scientists in writing. They do not train English majors—only science majors in how to write science. Artists of all kinds are being left behind in the Scientific Revolution simply because they do not have a basic knowledge of science.

Although commendable, MIT's efforts seem failed. Its *Leonardo* journal offers artistic efforts that look like science fair projects with boring representations of scientific principles with little artistic flare or grit. Few synthesize scientific data in a way that inspires. Their artists have a scientific vision that seemingly exists in some emotion-free, value-free, technical realm that rejects emotion, beauty, earthly human connection, and ethics.

Scholastic standards, overwrought prose, and false positives

Artists need higher scholastic standards. They need skills that prepare them to analyze their work based on criteria that includes the synthesis of scientific principles while inspiring the viewer. Natalie Jeremijenko is an environmental artist with a background in engineering, biochemistry, and environmental technology. She self-describes as both thinker and maker of things. She represents the future path for artists of the twenty-first century. Similarly, Aaron

Rosen interviewed contemporary artists to explore their intimate journeys of spiritual reflections.[1] He touts artists that reach for internal, spiritual energy wrenched from disciplined thought that sound within their own artistic process.

Much of current abstract art focuses on the internal, psychological journey of the artist, offering the viewer a subconscious stream of design within Gestalt-like, visual self-analysis. In and of itself, art created from a deep sense of self with an artistic eye stands as a valid point of view, but when it becomes an artist's entire body of artwork, it is not enough. It does not suffice as an ultimate end goal in and of itself. Art needs to grow beyond over-personalized thinking, and art schools need to refocus on a goal of substantive artistic training.

Overwrought prose that accompanies much art analysis must be rejected. An example of spurious, fluffed-up art-speak is Kastner's call for "action in a contextual orbit around a strongly gravitational cultural body."[2] What is a "contextual orbit" aside from pretentious nonsense intended to impress? What is a "gravitational cultural body"? It is not a sociological or anthropological concept. This hyperbolic affectation of literary astonishment is awful. When artists are asked to explain their work, such pseudo-sociological and psychological terms are accepted as if they were part of the art world's nomenclature and social structure, or worse, its market structure. Art critics, state your case and back it up with observation, evidence, and analysis. Include good photographs. Write simply and directly. Get to the point. Since when did this even become a question in the academic world?

The false positive is another all-too-common attitude. See for example *Weather Report: Art and Climate Change*[3] in which Lippard acted as curator for artists engaged to produce environmental art. Their works were poor, yet Lippard settled for all of it and even touts them in her book. Another example, Inigo-Manglano-Ovalle's *Iceberg Beyond the Irish Sea*, England, 2005, used data from a

1. Rosen, *Spiritual Traces*, xxi.
2. Kastner and Wallis, *Land and Environmental Art*, 17.
3. Lippard, *Weather Report.*

real iceberg to create his twenty-five-foot sculpture.[4] At best this work reflects only data and is outright boring without attempt to synthesize the data into an understanding of what it means. Another example is Natasha Mayers' *Whitefield Welcomes Global Warming/Whitefield Concerned Citizens Urge Action on Global Warming* (also entitled *Hot New Fashions from L. L. Bean Winter Collection*).[5] It is a picture of eight people waving and playing in winter outfits that have been cut, split, and dismembered to show more skin during hot weather. Juvenile. Another example is a hunter who is field dressing a caribou with (gasp) blood showing.[6] Again, juvenile, yet Mayer thinks this is great. It is not.

These curators settled for a pittance. Good mentors are not always nice. Sometimes people simply need to be told to do their jobs without excuses. I suggest that some curators may have had a secondary agenda: the more touted the artwork, the more kudos to the curator; the more visitors to the art show, the more money. Curators need to be held accountable, and likewise curators need to hold artists accountable. Remember that Michelangelo's papal mentor was not a simple, nice man that settled for so-so work. Michelangelo painted the Sistine Chapel because he had to, and he had to do a good job. False positive reviews of artists serve only to perpetuate poor work. Secondary agendas need to be substituted with demands for good, inspirational work.

On the opposite side of this spectrum, Mierle Laderman Ukeles exemplifies an artist striving to inspire others as she seeks a nexus between art and science. She began her work at a landfill by learning about landfill trash. Then she synthesized her knowledge into art and used it to inspire and serve the community. Her method brings synthesis to a functional level. While an area more often visited by engineers, it stands in the art world as a valid artistic endeavor.

Students should likewise be required to state the reasoning behind their artworks. Design alone is not enough. What is

4. Lippard, *Weather Report*, 81.

5. Lippard, *Weather Report*, 85.

6. Lippard, *Weather Report*, 25.

represented and why? What does it mean? What is the thought process behind it? How was it planned and executed, even if it was made from gunpowder like Cai Guo-Qiang?[7] Nothing less is acceptable in academia. Students should be pressed to synthesize their thoughts, experience, and knowledge with their artistic technique for the betterment of all and to inspire all. This should stand as the ultimate goal of art schools. Higher scholastic standards are needed for all aspects of the art world.

Substantive emphasis, or how to teach the Renaissance and Michelangelo

As we begin to emphasize context and substantive values in painting and visual artworks, we need to reassess how we teach students about the great artists of the Renaissance. For example, Michelangelo painted the universe and his vision of a Creator as he and the Roman Catholic Church understood it. Our commitment should be to paint the universe and its Creator (or no Creator) as we understand it. As humans reassess the universe, artists should join in with the same degree of thought and scientific detail as Michelangelo brought from the Bible, including all the direction he received from the Church. During the Renaissance, the earth was thought flat and the center of the solar system. God was conceived as an aging White male (like the Pope) rather than any conception of energy. The Renaissance generally reflected these values as set forth by the religious doctrine of the Catholic Church of its time.

Within our present golden age of astronomy, few artists have stepped up to reassess our perception of earthbound life and of the universe. In the last six-hundred-plus years, Renaissance art still stands as the best technical artwork, the most inspirational, and most outright beautiful art of the Western world. Renaissance art reflected the substantive values of its time. Today, artists need to emulate the Renaissance's emphasis on substance. While we may not agree with the message of these paintings, Renaissance art held

7. See Guggenheim Museum, "Cai Guo-Qiang." See also Amazon, "*Art21*."

substantive focus. Abstract art designs do not reflect a substantive basis; they are just pretty designs.

The Renaissance teaches us that substance counts. Michelangelo showed us that when an artist believes in the substance of what is created, the artist can reach down and express fundamental substance from within. It shows. It is great art. He carved his soul into marble, and it stands as genius even today.

Artists should be taught to explore global issues.

Art should enlighten, explore, reach for social impact, and confront ethical issues while being consistent with scientific knowledge and its issues. That is a big order, and it defines the great artist. Those on the forefront of life here on earth have addressed such issues for some time. Former President Bill Clinton's Conference on Global Initiatives tackled such issues every year such as development of women's roles, sustainability of resources, and the like. In an article for *Time* magazine, October 2012, Clinton said, "I firmly believe that progress changes consciousness, and when you change people's consciousness, then their awareness of what is possible changes as well—a virtuous circle."[8] When the Hubble Telescope tells us there are billions of galaxies, our conscious understanding of our true size in the universe must change. Michelangelo painted when people believed the world was flat and the sun revolved around the earth. Today, how do people grasp an internalized sense of the earth's minute size and the vastness of the universe? How does one conceive that the next galaxy is millions of light-years away? What is happening to our planet earth as oceans warm, coral dies, animals go extinct? There is little art that helps us conceive of all this other than the colorized images generated from the electronic data of astronomers. These issues are so overwhelming that they welcome denial. It is easier to turn one's head and not think of it. Artists and art schools must not turn

8. Clinton, "Case for Optimism," 38.

their heads while the most critical issues humans have ever faced stand smack in front of us.

Global issues are taught at the high school and undergraduate level across the country. An example of such education is the previously mentioned teaching aid entitled *Educating Globally Competent Citizens: A Tool Kit for Teaching Seven Revolutions.*[9] As a study and teaching guide, this "tool kit" (as it identifies itself) is structured based on its presentation of seven major global issues. It is a developed teaching tool that can challenge artists both creatively and scientifically to participate in global issues. It engenders substantive thought that can help artists begin to create visual images for viewers that support and provoke analysis. Why not require art students to participate in the Scientific Revolution by learning basic science and working with scientists in an effort to help us all build a new conscious image of what we are, what size we are, and where we are going?

Organization of art schools

Art schools have become the haven for those who act without discipline—the pampered, the lazy, the drug-laden, and generally anyone who colors themselves a "rebel." It emulates indulged adolescence in full bloom. Art schools tend to support the self-indulgence of prima donnas who consider mature artwork as something merely original and often without conceptual thought. Our culture imagines artists as unique individuals with outrageous hair or dress, as if an artist's life were itself performance art. Andy Warhol conjured such an image. Educational counselors tend to direct local crazies to art departments. The work is easy, you don't have to do much, and generally anything goes. It denigrates both art and the true scholarly business of education.

Unfortunately, the opposite extreme fails as well. Some art schools reach back to traditional artistic values such as hours and hours learning to draw realistically, making paints "the right" way

9. Falk et al., *Educating Globally Competent Citizens.*

(whatever that is), and learning portraiture, color wheels, and the like. I personally spent five classroom hours drawing toilet paper. It was humiliating and wrong again.

Art must become interdisciplinary training. Rather than compartmentalizing drawing, it may be combined with science classes. Better to spend five class periods drawing cells splitting or a virus invading rather than toilet paper. What about plant functions? What about galaxies? Science covers nearly every aspect of our lives today, and art should create with it, side by side. Art students need broad academic backgrounds to create from. 3D printers, lasers, and NASA photographs are three obvious examples.

Much abstract art seems to fall into the realm of *etudes* (studies). Like *etudes* in music to learn chords and scales, artists must practice to master their skills of drawing, painting, visual effects, and the like. For example, Rothko's huge gray panels present an excellent example of how dark tones are used and unfortunately also an example of untreated, severe depression. Agnes Martin painted what she called "the plain truth" with canvases that were only one, solid color or sometimes painted like a grid as if to compare scale. Her work is a study in simplicity as beauty. Pollock studied extreme paint application—beautiful in some ways, yet also a direct cause of indigestion. These studies have freed art from the ridged constraints of formal realism (still hanging around from the Renaissance), but they are studies. They are not great art. They lack substance. Great art has substance and holds its own valued reason and vision. As in music, *etudes* are not symphonies. Some *etudes* are good music, yet still not symphonies. Likewise, artistic studies of shapes and colors are valid to the extent they show us fundamental technique, but they are not great art.

The search for substance is the issue. Give students a basis for substance—science. Michelangelo painted the world as he knew it—flat and the center of the universe. In the twenty-first century, we must paint it as we know it, exploring life on earth within a universe larger and more mysterious than we ever imagined.

Art teachers generally have little background in science.

Art teachers were not hired for their background in science, and few have much scientific background. Like their students, they will need to retrain. More importantly, they must cooperate with the science departments of their schools. To be blunt, this foretells that for a period of time art will be in service of the science departments and perhaps lead by them. This means giving up precious power within academia and putting well-honed egos into hip pockets. Like their students, teachers need to have some means of taking science classes that does not call on them to calculate mysterious equations or face failure. Artists create because they are not scientists. I do not call for them to act and learn like scientists, but total ignorance of scientific issues must no longer be acceptable.

Art teachers are teachers and mentors that drive students to their highest level as visionaries.

Let us remember that Michelangelo was driven by a demanding Pope who did not give in to his whims and moods. Art teachers would do well to toughen up, both on themselves and their students. No excuses.

10

Implications for the Art Market and Art Museums

Art sales will grow within niche markets of scientific concern.

As art changes and begins to further realign on twenty-first-century issues, the message and content of each artwork will drive how and to whom it is sold. For example, those interested in specific content such as environmental issues will define a niche market, usually found online. The larger the niche with more people interested in a particular scientific area, the more art sales. The demographics of a niche will define its marketing strategies. For example, environmental concerns such as available water sources, riparian rights, and coral reefs would likely sell through environmental activist groups that meet online or read specific blogs or journals. Art with spiritual context would likely sell to those with similar interests such as progressive faiths, interfaith groups, at religious or group conferences, or meeting houses such as churches, temples, and mosques. Another example would be politically based work that would find a market among members of like-minded groups. As art focuses on content, it will be sold for its content and its degree of social impact rather than what it is made of, its style, or its technique. Did it help make a change? Were others affected by it? For example, recall the now-famous photograph of President Obama's face redrawn in red, white, and blue that became a well-recognized "face" of his presidential campaign.

Walking into a plush gallery will become the least effective and costliest means of making sales. Maintaining a storefront requires full-time staff as well as rent, insurance, utilities, etc. If the market is the extreme top 1 percent of income earners, then a gallery in New York or Hong Kong is wonderful. However, the art world can make a global impact though online sales of niche exhibits that reach specific demographics with less cost.

Art is now generally sold based on the reputation of the artist as held by art critics, peer reviews, gallery owners, museum curators, and the price of the artist's previous works. The buyer cries, "Oh! A Warhol," and its value soars. I suggest the value of reputation sales is artificially inflated by the art world itself to increase profits. Auction houses, galleries, buyers, and the artists themselves all want to make money. After all, if a collector owns several Andy Warhols that the collector bought as an investment, the collector is motivated to increase the value of those artworks regardless of whether his work has meaning or not.

Present focus on the reputation of individual artists has made some a near-personality cult, nurtured by romantic notions of the "rebel artist." For example, Mark Rothko, in real terms, was a manic-depressive who sadly killed himself. Some critics rather flippantly opine that his dark gray and black canvases demonstrate great technique in the use of subtle shades of gray that resonate with spiritual connection. It is an assumed attitude to increase sales and/or value of his work. The apparent truth lies more in their denial of Rothko's pain as he fell into a suicidal abyss. Similarly, there exists the charade that Jackson Pollock's alcoholic binges were "cool" or that throwing paint at a canvas was art worthy of national acclaim. The man was ravaged by alcohol and so was his art. Joan Mitchell was a scribbler. Willem de Kooning was a ghastly scribbler. To elevate their artwork to world-class is art without substance, like a king without clothes. If you wish pleasing designs of varied colors for your home, office, or bathroom, you are entitled, but do not call it great art. Great art has substance and addresses the issues of our time.

Follow the money—art museums

The art market is directly connected to art museums. Curators and museum directors, whether by design or mere compliance, artificially inflate prices to protect collections at their museums and the private collections of their donors. Wealthy donors sit on their boards and hiring committees. Few have insightful vision of where the art world should or might go, so the norm becomes that of evaluating meaningless designs and making sure they bring a good price.

Art museums are beginning to change. I offer my respect to the New York Museum of Art for presenting exhibits of substance, although most of historic interest, but now trending more toward a future vision. As art begins to focus on issues of the twenty-first century, the substantive content of exhibits and the works displayed will drive how it is displayed and to whom it is sold.

Conclusion

Do the Right Thing for the Right Reason.

MOST ARTISTS LACK DEVELOPED vision of why they paint. They may wish for an interesting design or a pretty picture, but does their work have meaning? Merely doing what feels good or what one thinks is right for your personal psyche is never enough. It never has been. It is one of life's great lessons that you must do both the right thing and for the right *reason*. Artists must have knowledge and vision and give thoughtful content to their expression.

We live on the cusp of the Scientific Revolution and welcome the golden age of astronomy. Earthbound philosophies and ethics with outmoded religious notions that violate our scientific knowledge are vacuous. Art for art's sake is dead. It lacks fundamental substance for the twenty-first century. It is self-absorbed. Our great museums curry favor of the *uber*-rich and suck from our society funds and concern that should serve more pressing needs.

Where do we come from? What are we? Where are we going?

This book challenges contemporary abstract artists to reach beyond abstract designs and grapple with the impact of twenty-first century science and its need for ethics, meaning, and substance while not abandoning their freedoms.

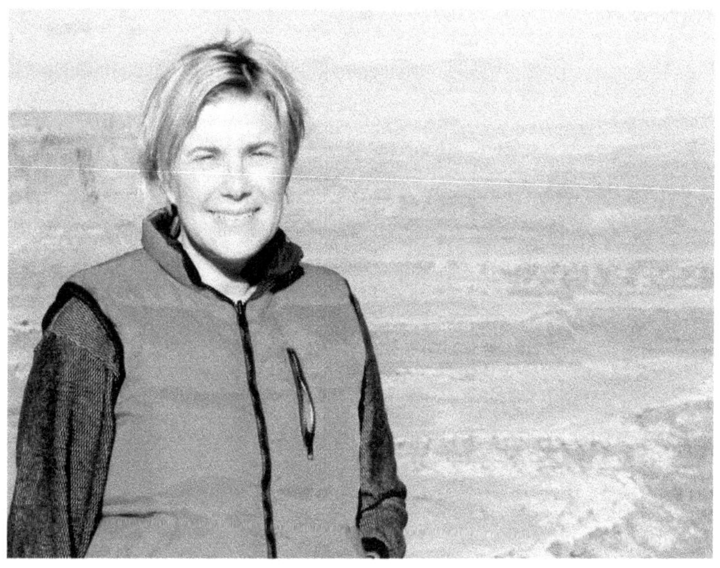

Elta at Canyonlands National Park—I was younger then! Contact me at eltawilson@gmail.com. I enjoy speaking with groups and can assist with curriculum discussion. All are welcome.

Curriculum Guide

Issues to Ponder

EACH CHAPTER IS LISTED with its subheading followed by discussion questions.

Introduction: Where do we come from? What are we? Where are we going?

Do you think the questions posed by Paul Gaugin in 1898 are relevant today? Do you think they direct us to a future vision of art in today's world?

Chapter 1: Vision for the Twenty-First Century: The Issue is Art About What and Why.

- The Twenty-First Century cries out for synthesis of scientific thought with the expression of ethics at its highest level, fostering a form of conceptual art at its ethical, philosophical best.
- The Scientific Revolution propels us into The Great Age of Discernment in direct challenge to artists.
- Artists must seek a nexus of art, science, and ethics—and shatter the images and idols of traditional thought that now should be guided by our understanding of the Universe.
- The central question is, "Art about what and why?"

Do you sense that the generation born within this century has experienced dramatic change created by science? If so, what are these changes and how have they affected you and/or your family and friends?

Do you find there are set or standard structures of traditional thought that need to be challenged because of these changes?

Do you agree/disagree with the book's premise that the issue is now, "Art about what and why?"

Chapter 2: The Scientific Revolution is Here Now, So Wake Up.

- The Scientific Revolution brings knowledge that births a New Matrix.
- A glance at a historical analysis.
- Envision for the Twenty-first century a New Matrix within a nexus of art, science, and ethics.
- Great art will synthesize questions that count.

Do you think that this century has brought such dramatic changes through new scientific knowledge that it may justifiably be called a "Scientific Revolution"?

What artworks in the 1900's foretells or envisions great artwork today?

What do you think constitutes "great art"?

Chapter 3: Scientists and Artists are Both Visionaries, and Both Must Heed the Role of Ethics.

- Without ethics, both science and art are meaningless.
  - — Science without ethics can be depraved.
  - — Science gives us humility as our compass.
  - — Artists are called to symbolic expression of self-reflection and substantive ethics.

- Both artists and scientists are called to envision a future through a disciplined inner life that contemplates the mysteries of the Universe and humanity's relationship with it.
- Art and science share in spiritual and ethical insight.
- It is this nexus of art, science, and ethics that poses the great conundrum of the Twenty-first century. While science grounds our understanding of the known Universe, our instincts continue to lead us, and our truth is heard by the soul.

Do you think that scientists and artists are both visionaries? Are you a visionary? If so, what do you foretell for the future of great artwork?

Do you agree with the author that humility is an artist's compass? Is it your compass?

How do you consider the role of ethics in your artwork?

Chapter 4: An Offer of Inspiration: NASA Images

Do you find NASA inspiring?

What inspires you in your artwork?

Chapter 5: A Plea for Artistic Content and Thought.

- Art must move beyond self-absorption.
- The historical approach to art analysis fails to confront rapid, scientific change and the need for ethical response.
- Examples of "Environmental Art" demonstrate a failure to find the nexus between art, science, and ethics.
- "Environmental Art" is really "Environmental Protection Art."
- Artists and scientists are both visionaries, and artists and art schools need the focus of substantive issues.

If artists and scientists are both visionaries, how can art studies begin to approach this issue and try to develop such vision in aspiring artists?

Chapter 6: Artists and Scientists Need Each Other.

Do you think scientists and artists need each other? How would such developed relationships add to an artist's work?

Chapter 7: Juvenile Narcissism as Art

Calling another artist's work out as "Juvenile Narcissism" is a strong, judgmental, label. Is it justified?

Chapter 8: Art for Art's Sake Is Dead—Substance Lives On.

- Art's role in the Virtuous Circle
- Freeform design is by definition arbitrary.
- Pure abstraction equates to pure artistic expression.
- Global issues and the artist
- Collaboration required.
- Great art calls to virtue and to science.

What is the basis, if any, for evaluating artwork that consists of only freeform design?

Does art make an ethical contribution to culture, meaning here any culture whether provincial, national, or global?

Do you believe your artwork can contribute to virtue or ethics of a viewer?

Chapter 9: Implications for Art Schools: The Search for Substance

- Scholastic standards and overwrought prose and false positives
- Substantive emphasis or how to teach the Renaissance and Michelangelo
- Artists should be taught to explore global issues.
- Organization of art schools
- Art teachers generally have little background in science.
- Art teachers are teachers and mentors that drive students to their highest level as visionaries.

How would you change art schools or art courses or art retreats or even art awards to help developing artists impact ethics and envision the dramatic changes fostered by what the author terms the Scientific Revolution?

Have you ever encountered an artist that was conversant or trained in an area of science? Did it impact your artwork?

Chapter 10: Implications for the Art Market and Art Museums

- Art sales will grow within niche markets of scientific concern.
- Follow the money—art museums!

How can an artist concerned with substantive issues confronting all areas of social, national, and global concern make a reasonable living? How can a visionary, ethical artist get paid?

Conclusion: Do the Right Thing for the Right Reason.

- Where do we come from? What are we? Where are we going?

Why do you engage with life as an artist?

Bibliography

Amazon. "*Art21: Art in the Twenty-First Century*." https://www.amazon.com/Art21-Art-Twenty-First-Century/dp/B002VV8M24.

Beardsley, John. *Earthworks and Beyond*. 4th ed. New York: Abbeville, 2006.

Cassidy, Harold Gomes. *The Sciences and the Arts: A New Alliance*. New York: Harper & Brothers, 1962.

Clinton, William. "The Case for Optimism." *Time*, October 1, 2012, 38–44.

Ede, Sian. *Art and Science*. New York: I. B. Tauris, 2005.

Einstein, Albert. "Interview with Albert Einstein: Death of a Genius." Interview by William Miller. *Life Magazine*, May 2, 1955, 61.

———. *The World As I See It*. California: Snowball, 2017.

Falk, Dennis, et al., eds. *Educating Globally Competent Citizens: A Tool Kit for Teaching Seven Revolutions*. Washington, DC: Center for Strategic International Studies, 2012.

Guggenheim Museum. "Cai Guo-Qiang Creates New Gunpowder Paintings." YouTube. December 13, 2019. https://www.youtube.com/watch?v=6al_eiTc67M.

Kaster, Jeffrey, and Brian Wallis, eds. *Land and Environmental Art*. New York: Phaidon, 1996.

Kleiner, Fred. *Gardner's Art Through the Ages: A Global History*. 13th ed. Boston: Thomson-Wadsworth, 2009.

Lippard, Lucy, curator. *Weather Report: Art and Climate Change*. Colorado: Boulder Museum of Contemporary Arts, 2007.

Massachusettes Institute of Technology, *MIT Leonardo Journal*, https://www.leonardo.info/.

NASA. "Images and Media." https://www.nasa.gov/nasa-brand-center/images-and-media/.

———. "NASA Image and Video Library." https://images.nasa.gov.

Rosen, Aaron. *Spiritual Traces: Reflections and Conversations on Contemporary Art*. Eugene, OR: Cascade, 2025.

Shiner, Larry. *The Invention of Art: A Cultural History*. Chicago, IL: The University of Chicago Press, 2001.

United States Space Force. "About Us." https://www.spaceforce.mil/About-Us/.

———. "FAQ." https://www.spaceforce.com/faq.

———. "Space Force Announces New Mission Statement." https://www.spaceforce.mil/News/Article-Display/Article/3517324/space-force-announces-new-mission-statement/.

Wilson, Stephen. *Art and Science*. New York: Thames and Hudson, 2010.

Index

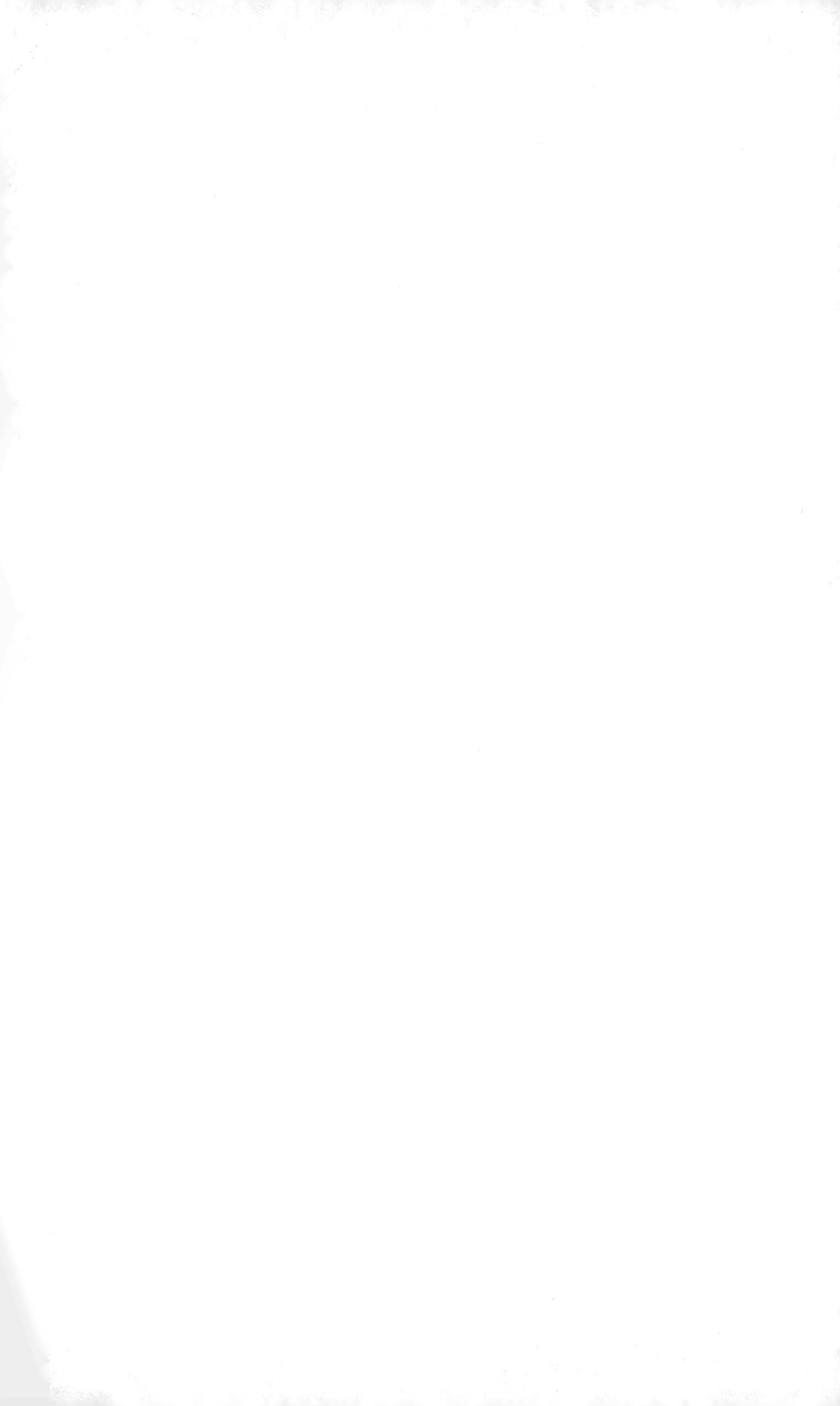

www.ingramcontent.com/pod-product-compliance
Lightning Source LLC
LaVergne TN
LVHW020657100826
845148LV00012B/2535